The Complete Guide to Survival Farming

Beginner's Guide to Growing Your Own Food

Jade Green

Table of Contents

Introduction

Imagine a world where you don't have to worry about where your next meal is coming from. A world where you can grow your own food, no matter what the circumstances. A world where you're self-sufficient and in control of your own destiny.

This is the world of survival farming.

Survival farming is the practice of growing your own food, even in the most difficult of circumstances. It's about being prepared for anything, from natural disasters to economic collapse. It's about taking control of your own food supply and ensuring that you and your family will always have something to eat.

If you're interested in learning more about survival farming, then this book is for you. Inside, you'll find everything you need to know to get started, from choosing the right location to harvesting your crops. You'll also learn about the benefits of survival farming, the challenges you'll face, and how to overcome them.

So what are you waiting for? Start your journey to self-sufficiency today!

Chapter 1

Why You Should Grow Your Own Food

In today's world, it is more important than ever to be self-sufficient. With the rising cost of food and the ever-increasing threat of natural disasters, it is essential to have a way to provide for yourself and your family. Survival farming is the practice of growing your own food, even in the most difficult of circumstances. It is a way to ensure that you will always have something to eat, no matter what happens.

There are many reasons why you should grow your own food yourself. Here are just a few:

It is cheaper than buying food from the store. The cost of food has been rising steadily for years, and it is only going to continue to go up. By growing your own food, you can save a significant amount of money.

It is healthier than food from the store. Food that is grown at home is not exposed to the same pesticides and herbicides as food that is grown commercially. This means that it is healthier for you and your family.

It is more sustainable. Growing your own food yourself is a more sustainable way to live. It reduces your reliance on the grocery store and helps to protect the environment.

It is a rewarding experience. Growing your own food yourself is a more rewarding experience. It is a great way to get some exercise done, fresh air, and sunshine. It is also a great way to teach your children about where the food they consume comes from.

The Benefits of Survival Farming

In addition to the reasons listed above, there are many other benefits to survival farming. Here are just a few:

It can help you to become more self-reliant. When you grow your own food, you are not relying on the grocery store or anyone else for your food supply. This can give you a sense of security and keep your mind at peace.

It can help you to learn new skills. Growing your own food is a great way to learn new skills, such as gardening, composting, and canning. These skills can be very valuable in a survival situation.

It can help you to connect with nature. Growing your own food by yourself is a great way to connect with nature. It can help you to appreciate the beauty of the world around you and the importance of taking care of the environment.

It can be a fun and rewarding hobby. Growing your own food by yourself can be a fun and rewarding hobby. It is a great way to get some exercise done, fresh air, and sunshine. It is also a great way to teach your children about where the food they consume comes from.

How to Get Started

If you are interested in getting started with survival farming, there are a few things you need to do. First, you need to choose a location for your garden. The location should have good sunlight, soil, and water. You also need to decide what kind of crops you want to grow. Once you have chosen a location and crops, you need to prepare the soil and plant your seeds. After that, you need to water, fertilize, and weed your garden. Finally, you need to harvest your crops when they are ripe.

Growing your own food can be a lot of work, but it is also a very rewarding experience. By following these tips, you can get started with survival farming and enjoy the many benefits it has to offer.

Chapter 2

Choosing the Right Location

The first step to starting a survival garden is to choose the right location. The ideal location will have:

Full sun: Most plants need a minimum of 6 hours of sunlight for their growth every day.
Good soil: The soil should be fertile and well-drained.
Access to water: You will need to water your garden regularly, so it is important to have access to a water source.
Protection from pests and diseases: If you live in an area with pests or diseases, you may need to take steps to protect your garden.

Soil

The soil is the foundation of your garden, so it is important to make sure it is in good condition. To improve the quality of your soil by adding compost, manure, or other organic matter. You should also test your soil to make sure it has the right pH level.

Sunlight

Most plants need at least 6 hours of sunlight every day for growth. If you do not have a spot in your yard that gets full sun, you can grow plants that tolerate partial shade.

Water

Water is essential for plant growth. You will need to water your garden regularly, especially during hot, dry weather. The amount of water you need to give your plants will vary depending on the type of plants you are growing and the climate you live in.

Pests and Diseases

No garden is immune to pests and diseases. However, there are important steps you can take to protect your plants. Pests and diseases can be prevented by doing the following:

Planting resistant varieties: There are many varieties of plants that are resistant to pests and diseases.
Practising good sanitation: Keep your garden clean and free of debris.

Inspecting your plants regularly: Check your plants for pests and diseases on a regular basis.

Using natural pest control methods: There are many natural ways to control pests and diseases. Some examples include:

Biological control: This involves using natural predators to control pests.

Insecticidal soap: This is a mild soap that can be used to kill insects.

Neem oil: This is an oil that can be used to control a variety of pests and diseases.

By following these tips, you can choose the right location for your survival garden and give your plants the best chance of success.

Chapter 3

Planning Your Garden

Once you have chosen a location for your garden, you need to start planning what you are going to grow. There are a few things to put in consideration when planning your garden:

What do you want to grow? This is the most important question to ask yourself. What kind of food do you and your family prefer to eat? Once you know what you want to grow, you can start researching different varieties of plants and choosing the ones that are best suited for your climate and growing conditions.

How much space do you have? This will determine the size of your garden. If you have a small space, you can grow vegetables in containers or raised beds. If you have a larger space, you can grow vegetables in the ground.

When do you want to harvest your crops? Some crops, such as tomatoes, take several months to mature. Others, such as lettuce, can be harvested in just a few weeks. Consider when you want to harvest your crops when planning your garden.

Crop Rotation

Crop rotation is the practice of planting different types of crops in the same area every year. This helps to keep the soil healthy and prevents pests and diseases from building up. There are a few different ways to rotate your crops. One way is to plant a different family of crops in the same area each year. For example, you could plant a legume crop (such as beans) one year, a cereal crop (such as corn) the next year, and a vegetable crop (such as tomatoes) the year after that. Another way to rotate your crops is to plant a different variety of the same crop in the same area each year. For example, you could plant a short-season tomato variety one year and a long-season tomato variety the next year.

Companion Planting

Companion planting is the practice of planting certain types of crops together. Some crops benefit from being planted together, while others can actually harm each other. There are a few different reasons why companion planting is beneficial. For example, some crops can help to attract beneficial insects, while others can help to repel pests. Some crops can also help to improve the soil quality.

Spacing

When planting your crops, it is important to space them properly. This will ensure that they have enough room to grow and develop properly. The amount of space you need to give each crop will vary depending on the type of crop and the size of your garden. You can find spacing recommendations for different crops in most gardening books or online resources.

Seed Selection

When choosing seeds, it is important to select varieties that are suited to your climate and growing conditions. You should also consider the maturity date of the variety. Some varieties mature early, while others mature late. Choose a variety that will mature before the first frost in your area.

By following these tips, you can plan your garden and give your plants the best chance of success.

Chapter 4

Soil Preparation

The soil is the foundation of your garden, so it is important to make sure it is in good condition. You can improve your soil quality by adding compost, manure, or other organic matter to it. You should also test your soil to make sure it has the right pH level.

Composting

Composting is the process of converting organic materials into nutrient-rich soil amendments. You can make your own compost by collecting food scraps, yard waste, and other organic materials in a compost bin or pile. Once the materials have decomposed, you can use the compost in your garden to improve the soil quality.

Tilling

Tilling is the process of turning over the soil with a shovel or tiller. This helps to aerate the soil and break up any clumps. Tilling is not always necessary, but it can be helpful if your soil is compacted or if you are planting in a new area.

Amendments

Soil amendments are materials that you add to the soil to improve its quality. Some common soil amendments include compost, manure, lime, and gypsum. You can add soil amendments to your garden at any time, but it is best to do so in the fall or early spring.

By following these tips, you can prepare your soil for planting and give your plants the best chance of success.

Here are some additional tips for soil preparation:

Add compost to your soil every year: Compost helps to improve the soil structure, drainage, and fertility.
Test your soil pH level and add lime or sulphur if necessary: The ideal pH level for most plants is between the value 6.0 and 7.0.
Amend your soil with organic matter: such as manure or compost, to improve its fertility.
Till your soil in the fall to a depth of 6 to 8 inches: This will help to aerate the soil and break up any clumps.
Rake your soil smooth before planting: This will help to create a level surface for your plants.

Chapter 5

Watering

Water is essential for plant growth. You will need to water your garden regularly, especially during hot, dry weather. The amount of water you need to give your plants will vary depending on the type of plants you are growing, the size of your garden, and the climate you live in.

The following are some tips for watering your garden:

Water deeply and less often: This will help your plants develop deep roots that can reach water even during dry periods.

Water early in the morning: This will help the water soak into the soil before the sun gets a chance to evaporate it.

Water at the base of the plants: This will help prevent the leaves from getting wet, which can promote fungal diseases.

Mulch your garden: Mulch helps to retain moisture in the soil and in suppressing weeds.

Irrigation

There are a number of different ways to irrigate your garden. Some common methods include:

Hand watering: This is the most basic method of irrigation. You can water your garden by hand with a hose, watering can, or bucket.

Soaker hoses: Soaker hoses are long, flexible hoses that are filled with water. They release water slowly and evenly, which helps to prevent water runoff and evaporation.

Drip irrigation: Drip irrigation systems use a network of small pipes to deliver water directly to the roots of your plants. This is a very efficient way to water your garden, as it minimises water waste.

Micro-sprinklers: Micro-sprinklers are small, rotating sprinklers that deliver water in a fine mist. They are a good choice for watering small areas, such as flower beds and vegetable gardens.

The best method of irrigation for your garden will depend on the size of your garden, the type of plants you are growing, and your budget.

Mulching

Mulch is a layer of material that you spread over the soil in your garden. Mulch helps to retain moisture in the soil, suppress weeds, and improve the overall health of your plants.

There are a number of different materials that you can use as mulch, including:

Wood chips: Wood chips are a popular choice for mulch. They are relatively inexpensive and very easy to find.

Compost: Compost is a great way to recycle organic materials and improve the soil quality in your garden.

Grass clippings: Grass clippings can be used as mulch, but they should be applied in a thin layer and allowed to dry before they are spread.

Straw: Straw is a good choice for mulch in vegetable gardens. It helps in keeping the soil cool and moistened.

Mulch should be applied to your garden in a layer that is 2 to 4 inches thick. You will need to reapply mulch as needed throughout the growing season.

Chapter 6

Fertilizing

Fertilizer is a substance that provides nutrients to nourish the plants. It can be organic or synthetic. Organic fertilizer is made from plant or animal materials, while synthetic fertilizer is made from chemicals.

There are many different types of fertilizer available, and the best type for your garden will depend on the type of plants you are growing, the soil conditions in your garden, and your budget.

The following are some tips for fertilizing your garden:

Test your soil: Before you fertilize your garden, it is important to test your soil to see what nutrients it needs. You can do this by sending a sample of your soil to a lab or by using a home soil test kit.

Use the right type of fertilizer: Once you know what nutrients your soil needs, you can choose the right type of fertilizer. Organic fertilizers are a good choice for gardens that are not heavily fertilized. Synthetic fertilizers are a good choice for gardens that need a lot of nutrients.

Fertilize at the right time: It is important to fertilize your garden at the right time. Most plants need to be fertilized in the spring and fall.

Fertilize correctly: When you fertilize your garden, it is important to follow the directions on the fertilizer label. Over-fertilizing can damage your plants.

Organic Fertilizers

Organic fertilizers are produced from plant or animal materials. They are a good choice for gardens that are not heavily fertilized. Organic fertilizers release nutrients slowly, which helps to prevent nutrient burn.

Some common types of organic fertilizers include:

Compost: Compost is made from decomposed organic materials, such as food scraps, yard waste, and manure.

Manure: Manure is a good source of nitrogen, phosphorus, and potassium.

Blood meal: Blood meal is a good source of nitrogen.

Fish meal: Fish meal is a good source of nitrogen and phosphorus.

Bone meal: Bone meal is a good source of phosphorus.

Synthetic Fertilizers

Synthetic fertilizers are made from chemicals. They are a good choice for gardens that need a lot of nutrients. Synthetic fertilizers release nutrients quickly, which can be helpful for plants that need a lot of nutrients to grow quickly.

Some common types of synthetic fertilizers include:

Nitrogen fertilizer: Nitrogen fertilizer is a good source of nitrogen.
Phosphorus fertilizer: Phosphorus fertilizer is a good source of phosphorus.
Potassium fertilizer: Potassium fertilizer is a good source of potassium.

Which type of fertilizer is right for you?

The best type of fertilizer for you will depend on the type of plants you are growing, the soil conditions in your garden, and your budget. If you are not sure which type of fertilizer is right for you, it is always best to consult with a gardening expert.

Chapter 7

Pest Control

Pests can be a major problem in gardens, and they can damage plants, spread diseases, and make it difficult to enjoy your outdoor space. There are a number of different ways to control pests, including natural and chemical methods.

Natural Pest Control

Natural pest control methods are a good choice for people who want to avoid using chemicals in their garden. There are a number of different natural methods that can be used to control pests, including:

Trapping: Traps can be used to catch pests, such as mice, rats, and insects. There are a number of different types of traps available, and the best type for you will depend on the type of pest you are trying to control.

Pest-repelling plants: There are a number of plants that can be used to repel pests. Some common pest-repelling plants include marigolds, citronella, and lavender.

Biological control: Biological control involves using natural predators to control pests. Some common biological control agents include ladybugs, praying mantises, and nematodes.

Habitat modification: Habitat modification involves making changes to your garden that will make it less attractive to pests. Some common habitat modification techniques include removing food sources, providing water, and creating shelter for beneficial insects.

Chemical Pest Control

Chemical pest control methods can be effective in controlling pests, but they should be used as a last resort. Chemical pesticides can be harmful to the environment and to people, so it is important to use them carefully.

When using chemical pesticides, it is important to read the label carefully and follow the instructions. You should also wear gloves and other protective gear when handling pesticides.

Which type of pest control is the best for your garden?

The best type of pest control for you will depend on the type of pest you are trying to control, the size of your garden, and your budget. If you are not sure which type of pest control is right for you, it is always best to consult with a gardening expert.

Chapter 8

When to Harvest

The best time to harvest your vegetables will vary depending on the type of vegetable you are growing. Some vegetables, such as tomatoes, can be harvested over a period of time, while others, such as lettuce, should be harvested all at once.

The following are some tips for knowing when to harvest your vegetables:

Look for signs of ripeness: Most vegetables will have some sort of sign that they are ripe. For example, tomatoes will turn red when they are ripe, and cucumbers will have a hard skin when they are ripe.

Taste test: If you are not sure if a vegetable is ripe, you can always taste it. Most vegetables will taste best when they are ripe.

Harvest early: If you harvest your vegetables too early, they may not be ripe enough to eat. However, if you harvest them too late, they may become overripe and mushy.

Chapter 9

How to Harvest

The best way to harvest your vegetables will also vary depending on the type of vegetable you are growing. Some vegetables, such as tomatoes, can be picked by hand, while others, such as lettuce, need to be cut with a knife.

The following are some tips for harvesting your vegetables:

Use sharp tools: When harvesting your vegetables, it is important to use sharp tools. This will help to prevent damage to the vegetables.

Harvest cleanly: When harvesting your vegetables, it is important to harvest them cleanly. This means cutting the vegetables at the base of the stem or removing them from the ground without damaging the roots.

Handle gently: When harvesting your vegetables, it is important to handle them gently. This will help in preventing bruising and damage.

By following these tips, you can harvest your vegetables at the peak of ripeness and ensure that they are of the best quality.

Chapter 10

Storing Your Harvest

There are a number of different ways to store your harvest, and the best method for you will depend on the type of vegetable you are storing and how long you need to store it.

The following are some tips for storing your harvest:

Cool, dark place: Most vegetables will store best in a cool, dark place. The ideal temperature for storing vegetables is between 35 and 40 degrees Fahrenheit.

Moisture: Vegetables need some moisture to store well. If the air is too dry, the vegetables will dry out. If the air is too humid, the vegetables will rot.

Avoid bruising: Bruised vegetables will spoil more quickly. Handle them with care and avoid them dropping.

Eat or freeze promptly: Some vegetables, such as lettuce, should be eaten or frozen as soon as possible after harvest. Others, such as carrots, can be stored for several weeks in a cool, dark place.

Canning

Canning is a method of food preservation by sealing it in an airtight container. This prevents the growth of bacteria and mould, which can cause food to spoil.

Canning can be done at home with a pressure canner or a water bath canner. It is important to follow the instructions carefully when canning, as improper canning can lead to foodborne illness.

Freezing

Freezing is another method of preserving food. It is a quick and easy way to store food for long periods of time.

When freezing food, it is important to blanch it first. Blanching helps to preserve the colour, flavour, and texture of the food. It also kills off any bacteria that may be present in the food.

Once the food has been blanched, it can be frozen in airtight containers or bags. It is important to label the containers or bags with the date so that you know when the food was frozen.

Freezing can be done at home or in a commercial freezer. It is important to follow the instructions carefully when freezing food, as improper freezing can lead to food spoilage.

Survival farming is the practice of growing food in a way that is sustainable and resilient to shocks and stresses, such as climate change, economic instability, and political upheaval. It is a growing movement, as people become increasingly aware of the need to be self-sufficient and to protect their food supply.

There are many different ways to practise survival farming. Some people choose to grow their own food in small gardens, while others choose to start their own farms. There are also a number of organisations that offer training and support for survival farmers.

The future of survival farming is bright. As the world becomes more unstable, people will increasingly need to be able to grow their own food. Survival farming is a sustainable and resilient way to grow food, and it is a skill that everyone should have.

Here are some of the benefits of survival farming:

- It is a way to ensure that you have access to food, even if there is a disruption to the food supply.
- It is a way to reduce your reliance on the global food system.
- It is a way to connect with nature and to learn about sustainable agriculture.
- It is a way to get exercise and to be more self-sufficient.
- It is a way to connect with your community and to build a more resilient food system.

If you are interested in learning more about survival farming, there are a number of resources available. You can find books, websites, and organisations that can provide you with the information and support you need to get started.

www.ingramcontent.com/pod-product-compliance
Lightning Source LLC
Chambersburg PA
CBHW070336240526
45466CB00027B/2105

9 7 9 8 3 9 4 4 9 2 1 6 7